# Practical small scale queen rearing using the Miller method

Lynfa Davies, Master beekeeper, NDB

Northern Bee Books

Practical small scale queen rearing using the Miller method
© Lynfa Davies 2022

First published 2022
Published in the United Kingdom by
Northern Bee Books,
Scout Bottom Farm,
Mytholmroyd,
West Yorkshire HX7 5JS
Tel: 01422 882751
Fax: 01422 886157
www.northernbeebooks.co.uk

ISBN 978-1-914934-26-1

Design and artwork, DM Design and Print

# Introduction

Rearing new queens for our colonies is a rewarding and important practice that often gets mis-labelled as too difficult and complicated. This, is simply, not correct and with some planning and attention to detail new queens can be produced at a fraction of the cost that you will pay for them elsewhere.

Queen rearing is an organised process where thought goes into selecting which colonies are used. It is not the same as producing new queens by using swarm cells from any colony that happens to be showing signs of swarming. When you select which colonies you want to raise queens from you are beginning the process of improving your stock. You can choose the colonies with the traits you like and avoid using those with traits that you do not like. In just a few seasons you can make a noticeable difference to your colonies.

Every year thousands of queens are imported into the UK to satisfy the demand from both commercial and hobby beekeepers. The reason for this is that countries in Europe (and beyond) can produce queens earlier in the season than we can in the UK and the spring is when the highest demand for queens occurs as beekeepers are sorting out the problems they find at the end of the winter and preparing for the forthcoming summer. However, with a little preparation and forward planning queens can be produced in the preceding season to accommodate demand the following spring. This means we can focus our attention on locally bred queens that are suited to our area and reduce the risk of importing pests and diseases. Rearing your own replacement queens is both interesting and fun but apart from that it makes sense financially. The small investment that may be required is quickly recouped once you have produced just a couple of new queens.

Fig 1. Raising your own queens is fun and cost effective!

# What is queen rearing?

Put simply, queen rearing is about using the natural mechanisms within a colony of bees to encourage them to produce queen cells. By manipulating that colony, they take the queen cells to maturity and then the beekeeper needs to intercept at the correct stage and move the cells to a small colony for the mating part of the process. There are many different methods of rearing new queens which is, most likely, why the process gets labelled as complicated. This booklet will focus on one method, the Miller method. This is a low-cost method because the beekeeper will already have all the equipment that is required. The production of the queen cells is relatively straightforward, and in many ways, is the easy bit. It is also important to pay attention to what you are going to do with these queen cells and how you will get the new virgin queens mated.

It is important to be realistic about the level of success you can expect. There are many points along the process where things can, and will, go wrong. Not all these things will be your fault and so to compensate, you need to plan on rearing more queens than you need. In the early days, if you require two new queens, aim to produce four and then at best you have a choice as to which ones to keep.

# Terms used in this booklet

Let us start with a brief explanation of some of the terms used in this booklet.

**Cell raiser colony** – this is the colony that you are using to produce queen cells from the young larvae that you introduce on the Miller frame. It needs to be a strong colony with plenty of young nurse bees. It also needs to be queenless so that the workers respond to this situation by readily producing new queens. There are situations where queenright colonies are used to produce queen cells but to avoid confusion this is not going to be dealt with in this book.

**Comb** – this refers to the wax comb that the bees build to rear their brood in and for the storage of honey.

**Donor colony** – this is the colony that you select as the one you would like to rear your queens from. This colony will provide (donate) the frame with eggs and young larvae that you will use to create your Miller frame. This colony will possess all the good traits that you want to encourage in your colonies such as good temperament, high health status and good honey yields.

**Drawn comb** – when bees are building their wax combs we call it 'drawing comb'. When they have finished building it the cells are back to back and deep enough to protect the developing larvae. This is called 'drawn comb'.

**Dummy board** – this is a wooden board, the size of a brood frame, that is used to fill any spaces in the brood box. It can act as a false wall which means a full-sized brood box can be used to accommodate smaller colonies by using a dummy board as the false wall. This means the colony does not have to try and heat an area that is too large for it.

Fig 2. Dummy boards can be used as false walls to change the size of the box.

**Egg** – these are the beginnings of all bees! Queens lay eggs into the wax cells and the workers look after them, keeping them warm and then feeding the larvae when they hatch. The eggs are long and thin and often said to resemble a candle wick sticking up from the bottom of the cell.

**Emergency queen cells** – when a colony finds itself in the situation of not having a queen it will take steps to produce a new one. To do this, the colony needs to have eggs or larvae that are less than five days old. If the emergency queen cell is made from a cell that contained a larvae that had started out developing as a worker then the shape of the cell takes on the characteristic elephant trunk shape. This is because the larvae that started out as a worker was in a horizontal cell. The colony then, for whatever reason, lost its queen, and the workers had to use these worker larvae and change the shape of the cell to accommodate the developing queen. Emergency queen cells tend to be on the face of the comb wherever the youngest larvae were in the brood nest and typically there may be 10 – 20 cells produced.

**Larva** – Three days after being laid the egg hatches and a larva emerges. These tiny grubs are fed by the workers and rapidly grow to become a pearly white, segmented larva that almost completely fills the cell. After spending five days as a larva gorging on food it then spins a cocoon and pupates before emerging as an adult bee.

**Mating nuc** – when you have multiple ripe queen cells you need to move them to mating nucs so that they do not all emerge together in your cell raiser colony. These nucs are made up of just a few frames of bees, brood and stores. The queen cell is introduced just before it is ready to emerge and the small colony will nurture her and then encourage her to go out on mating flights.

**Mini mating nuc** – a more efficient way of getting your new queens mated is to use mini mating nucs. This is useful if you have lots of queen cells because overall you will require fewer bees to make up the nucs. They are small boxes, usually polystyrene, with small frames in them that are not transferrable to your full size hives and nuc boxes. They are an ideal size for getting your new queens mated but are not designed for long term housing of a colony because once the queen is laying they will very quickly outgrow the box.

Fig 3. An example of a mini mating nuc.

**Queen cells** – These are the cells that the worker bees produce to protect the developing queens. They hang vertically from the comb and are cylindrical in shape. The workers start out by producing cups, sometimes called play cells. Only when an egg is laid into that cell do the workers start extending the length of it to accommodate the developing queen.

Fig 4. A queen cell on the edge of the comb.

**Queenless** – this is the term used when your colony does not have a queen. There are several reasons why this might occur and the situation can arise from natural causes or be induced by the beekeeper. For example, when setting up the cell raiser colony the queen is removed to create a situation that encourages the bees to make queen cells.

**Royal jelly** – when producing queens the workers feed the developing queen larvae with royal jelly. It is white in colour and the consistency of yoghurt. It is highly nutritious and provides the developing larvae with all the protein, energy and lipids that they require.

**Stores** – beekeepers refer to any honey in the hive as stores. It is important to be able to assess the amount of honey that is stored in the hive to determine whether the colony can continue to operate if there is no nectar flow. In this context, we need to be reassured that the cell raiser colony has plenty of stores to feed the workers that are producing the royal jelly even if there is a break in the nectar flow.

**Virgin queen** – when the new queen emerges from her cell and before she goes out on a mating flight she is referred to as a virgin queen. She is only likely to remain in this state for one to two weeks before the workers force her out of the hive to go on a mating flight.

# A refresher on the development stages of a queen honey bee

Successful queen rearing hinges around knowing what happens when in the development of the new queen so a quick refresher may be helpful.

From when the egg is laid to the new queen emerging takes only sixteen days and in this short time frame the incredible processes of development, pupation and maturation take place. It is vital that we know what stage of the process we are at, so that nothing is left to guess work. If we get it wrong, it is highly likely that multiple queens will emerge into the colony, and we risk losing them. This is a bit of a disaster after all our efforts to set up the process.

The queen will lay an egg on day one. This egg will be standing upright in the base of the cell. It is not easy to see but it is important that you practice looking for eggs. A torch or magnifying glass may help. They are much easier to see in fresh, clean comb rather than old, dark comb.

Between days one and three, the egg gradually moves in position until it is lying flat against the base of the cell. Remember the positioning of the queen cell. They hang down vertically from the comb and not horizontally like the worker and drone brood. So, while we describe the positioning of the egg as being in the base of the cell, that is as we look down into it, in reality, the egg is pinned to the top of the cell. Use the orientation of the egg to help you identify its age. If it is still upright, it is freshly laid and if it is lying down it will be close to hatching.

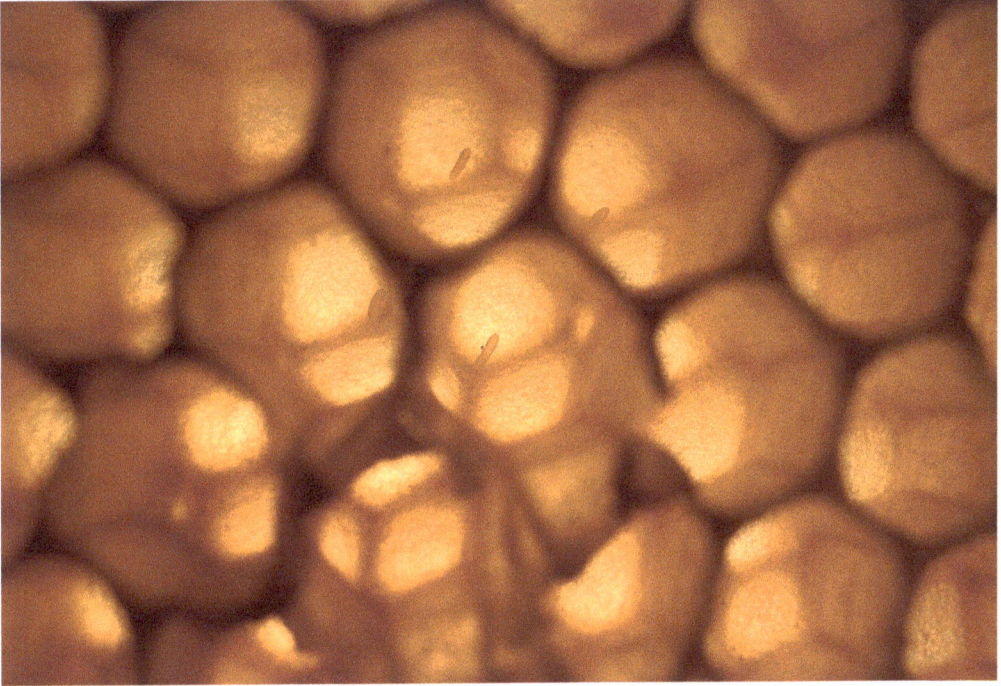

Fig 5. Eggs in the base of worker brood cells. Eggs that develop into queens look the same.

The egg hatches on day three to reveal a small, white larva. The workers begin feeding it and it will be lying in a pool of royal jelly. They continue feeding it over the next few days and it will grow rapidly to become a plump, segmented larva. This is the stage of development when we need to ensure the colony's nutrition is not interrupted. They need plenty of fresh pollen to make the royal jelly which will influence the quality of the queen produced. On day eight the larva will almost fill the cell and the workers seal it in ready for pupation. During this phase she will undergo several moults and eventually emerge as an adult on day sixteen.

**The critical stages for the queen rearer**

With the Miller method, and other methods of queen rearing, young larvae are transferred from the donor colony (that is the colony that you want to rear new queens from) into the colony that will rear them for you. The stage at which these larvae are transferred is important. They must be as young as possible and with the Miller method, that means either still as eggs or as larvae that are 24 hours old or less. The reason for this is, at the very early stages of development when the larvae are feeding in their cells, if they are to become a queen, the workers feed them excessive amounts of protein-rich royal jelly. They need to feed on as much of this as possible to switch on the genes that turn them into a queen rather than a worker, and to ensure their growth and development is as strong as possible so that the resulting queens are healthy and have large ovaries within their abdomens.

Another useful reference point is the day the queen cell is sealed. This happens on day eight and from this point you have eight days until the new queen emerges. You do not want to be opening the colony every day to see what stage the cells are at but as an example, if you put your Miller frame into the cell raiser colony and then come back three days later for a quick check what you should see is large, charged queen cells. If those queen cells are already sealed, then the larvae in the Miller frame were a little bit too old when they were introduced to the cell raiser colony.

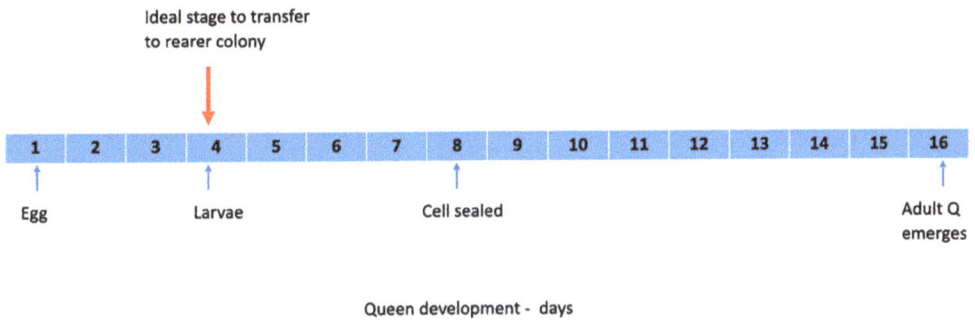

Ideal stage to transfer
to rearer colony

| 1 | 2 | 3 | 4 | 5 | 6 | 7 | 8 | 9 | 10 | 11 | 12 | 13 | 14 | 15 | 16 |

Egg                Larvae              Cell sealed                          Adult Q
                                                                           emerges

Queen development - days

Fig 6. The development stages of the queen.

Finally, the queen cell must be moved to where you want the queen to emerge before day sixteen. However, it must not be done too early because it is important that the developing queen is maintained at the correct temperature. If the cell is moved too early to a small mating nuc, the queen's development could be impacted or at worst, she may die. Timing this to take place on day fourteen at the earliest relies on you knowing the exact age of the queen cell.

# What do you need?

The donor colony

You will need a strong, productive colony that possesses qualities that you are happy to replicate. We'll call this the donor colony. When selecting your donor colony the best advice is to choose your best colony. When we are trying to improve the bees we keep, the aim should be to select from the top third of your colonies and requeen the bottom third. Then, over a period of just a couple of years, you can dramatically improve your stocks by selecting queens that produce colonies that are productive, healthy and overwinter well and avoiding ones where the colonies are aggressive and swarmy. The qualities you choose is up to you and some simple records can help you monitor these qualities.

Fig 7. Choose a strong productive colony with the traits you are looking for.

The cell raiser colony

You will also need a second colony that will become your cell raiser colony. This colony needs to be strong (covering 8-10 frames), have brood of all stages (because you need plenty of young bees) but doesn't need qualities that you particularly like because these won't get passed on to your new queen. The cell raiser colony is there to turn the eggs from your selected queen into queen cells by feeding the larvae with lots of high quality royal jelly. You could use your least favourite colony and then as part of the process allow them to keep a new queen that you rear! Honey production will be compromised in this colony because they will be without their queen for at least three weeks during the peak season.

Mating nucs

The third thing to consider is how you are going to get your new queens mated. The best way to do this is using a mating nucleus. This is a quantity of bees put into a nuc box or a mini mating nuc to look after the virgin queen until she gets mated. If you choose to go down the route of using mini mating nucs you will need to buy a few of these (depending on how many queens you intend to rear at any one time) and this is likely to be your only expense. The option that will be described here is to use a full size nuc box and again you may need to purchase one or two if you don't already have any. The downside of using full size nucs is you need more bees but if you are only planning to rear one or two queens that will be less of a problem. If you want to rear more queens than this then mini mating nucs are probably your best option.

# Working together

If you only have a couple of hives you may be thinking that queen rearing is not for you because it is possible that you do not have everything listed above available to you. This is where working together in pairs or small groups can really help. Between you, identify the hive with the best qualities to be the donor hive and another person can provide the cell raiser colony. Then between all of you, you can supply the bees for the mating nucs. Not only is this an efficient way to work but it can also be more successful as you bounce ideas off each other and work through the process together. Some of the timings described are also quite precise so having someone else in the system can help to ensure these timings are met.

Fig 8. Find a beekeeping buddy or buddies when you are queen rearing. It can make the process easier and more enjoyable.

# Preparation

Queen rearing is not something to start on the spur of the moment. A little planning will help enormously so that everything is in place when you need it. Firstly, don't start too early in the season. May is probably the earliest you can start and in some parts of the UK you may even need to leave it until late May or early June. You need your colonies to be strong so that they can produce plenty of nurse bees to care for the developing queen larvae. Your colonies, and those in the surrounding locality, need to have plenty of drones to mate with the new queens. But perhaps most importantly, you need the weather to be as reliable as possible – something that is hard to judge in the UK!! Settled weather helps on several counts; it will aid the development of your colonies in the spring; it encourages foraging that will provide all the nutrition your developing larvae need; it will help the small mating nucs to regulate their temperature and it will enable virgin queens to get out on their mating flights.

Preparation of the donor colony does not require any special intervention. Wait until they are a filling the brood box and by the time you start queen rearing it is likely they will have a couple of supers on as well. It helps if some of the frames in the brood box are fresh, newly drawn comb just because it is easier to see what you are doing with these frames than it is if the frames are old, dark comb.

The cell raiser colony needs to be prepared one week before you are ready to start your queen rearing process. Choose a strong colony and make it queenless by taking the queen out together with a couple of frames of brood and making up a nuc. After removing these frames from the colony, close up the gap and then add a couple of new frames (foundation or drawn comb) at the outer edge of the brood box. There is no need to place these frames in the middle because they don't have a laying queen to take advantage of them. In the nuc, make sure the queen has some stores as well as some space to lay and shake a couple of frames worth of extra bees to provide her with foragers. This nuc can be reunited with the cell raiser colony after the whole process is over.

Fig 9. Add a frame of pollen to the centre of the brood box to provide nutrition for the nurse bees feeding the larvae.

Leave the cell raiser colony for 5-7 days and when you come back they will have made emergency queen cells in response to the fact you made it queenless. You now need to remove these queen cells making sure you get every one. They are now in a situation that they cannot make more queen cells because they do not have larvae of the correct age to do this which means when you add the larvae from your donor colony they will be eager to draw queen cells for you. Keep an eye on this cell raiser colony. They need to have plenty of nectar and pollen coming into the hive to support the nurse bees who will be feeding the larvae you add. You may need to add a frame filled with pollen from another hive and place it next to the position where you will add the frame of larvae from the donor hive i.e. in the middle of the brood box. If the nectar supply slows down, or the weather turns unexpectedly cool just as you are about to add the frame of larvae from the donor hive, give them a small feed of syrup.

We will return to the preparation of the mating nucs later as they are not required just yet.

# The Miller method of raising queens

Once you have prepared your cell raiser colony and they have been queenless for a week you are ready to select a frame of suitable young larvae from your donor colony. The useful thing about the Miller method is you do not need any specialist equipment as you are going to get the cell raiser colony to make queen cells on a normal brood frame that you have taken from the donor colony.

Using fresh drawn comb is helpful because you can see the eggs and larvae easier and one way to make sure you have eggs on these frames at the time you want them is to introduce the frames into the centre of the donor colony's brood nest four days before you need them. This will provide the queen with new space to lay and hopefully provide you with a patch of eggs and newly hatched larvae on these frames. You are aiming to provide the cell raiser colony with larvae that are less than 24 hours old and they will use these larvae to make queen cells by feeding them large amounts of high quality royal jelly.

After four days select one of these frames with eggs or very young larvae on them and cut the comb so that the young larvae (or eggs as these will soon hatch to become young larvae) are along the cut edge. To prevent them from making clusters of queen cells that are difficult to separate pinch out some of the larvae along the edge to restrict the number they can make.

Fig 10. Locate the youngest larvae and make a zig zag cut where these larvae are positioned.

Fig 11. Use scissors to snip the wires.

Fig 12. The cut frame with position of larvae of correct age indicated.

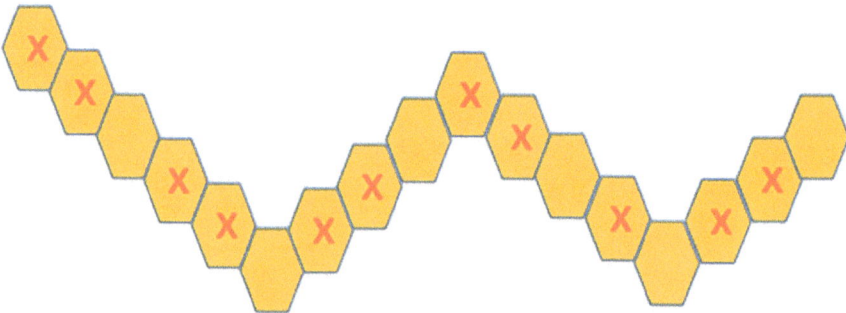

Fig 13. The red crosses represent the cells that have been pinched out along the edge to prevent them making queen cells that are clustered together.

Place this frame into the middle of the cell raiser colony immediately next to the frame of pollen. If you have selected very young larvae less than 24 hours old it will be another 4 -5 days before the queen cells they make are sealed so check 5-7 days after putting the frame in the cell raiser colony to see how many queen cells you have. Once you know how many queen cells you have you know how many mating nucs to prepare. Don't feel obliged to use every queen cell, you can give some away if you have more than you require. Also don't expect 100% success so even if you only want one new queen it is worth using two queen cells as an insurance policy. If they both result in mated queens then you can give one away!

This is where the timings get more critical. You need to transfer the queen cells to their mating nucs before they all emerge in the cell raiser colony. This is best done about two days before they are due to emerge but don't be tempted to do it too early as the pupating queens are very delicate and can be damaged if they get chilled. The picture (Fig 14) shows how queen cells have been produced along the cut edges of the Miller frame. These can be cut out using a knife but do make sure not to get too close with the knife and also take plenty of wax at the top to help you wedge it into place in the mating nuc. Cutting the queen cells from the Miller frame can be a little tricky as the bees often make multiple cells that are quite close together and that is why pinching out some of the larvae can help to reduce this. The bees may also replace some of the comb that was cut away and some of the cells become partly embedded in the comb. Great care is required when cutting the cells out and it is better to take too much comb than not enough. For this reason, using full sized nuc boxes is helpful because you have more space between the frames to pin a large lump of comb that contains your queen cell. In a mini mating nuc you have less space and then you are tempted to trim away the excess comb which can lead to damage.

Fig 14. Queen cells have been neatly made along the cut edge and these will be easy to cut out using a knife or scissors.

Fig 15. Carefully cut out the cell.

Fig 16. Leave plenty of wax around the cell to avoid damaging it.

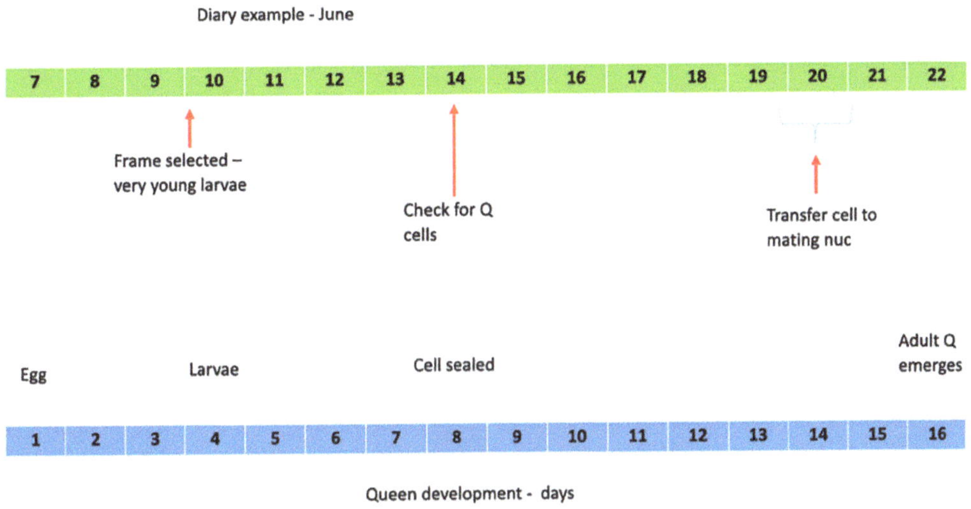

Diary example - June

Fig 17. The dates along the top are example dates in June and the numbers along the bottom show the number of days it takes a queen to develop from egg to emergence. In the example, an egg that is laid on the 7 June will be at the correct stage of development on your Miller frame on the 10 June. The queen cells will be sealed on the 14 June and then will emerge on the 22 June. This means they must be transferred to the mating nuc on the 20 or 21 June.

To complete this scenario the cell raiser colony would have been made queenless on the 3 June and then any emergency queen cells that they have made would be removed on the 10 June immediately prior to adding the Miller frame.

# Preparing your mating nuc

We will assume that you are only rearing a couple of new queens and in this scenario preparing two 2-3 frame nucs on full size brood frames will work well. You do not need full size nucs (i.e. 5-6 frames) because not only does this use a lot more bees and potentially depletes your other hives but small nucs work equally well to act as support system while your queen gets mated. You could of course, leave one queen cell in the cell raiser colony and allow them to requeen themselves with one of your selected queens and then you only need to make up one other mating nuc. If either are unsuccessful remember you still have the queen from the cell raiser colony in a nuc box as an insurance policy!

Fig 18. A full sized 6 frame nuc is not necessary to get your queen mated. A 2 or 3 frame nuc will suffice and use dummy boards to reduce the size of the box.

Choose a hive to steal some frames of brood and some bees from. If you have a limited number of hives available to you then this is where it helps to be working in a small group. Sacrifice your worst hive because once again you do not pass the traits of that hive to the new queen, they are just babysitting her. Now if you are making up two nucs in this way you need at least four frames of brood plus additional bees to shake into the nucs as well. These can either come from one colony which will then become severely depleted or if you have enough hives you can take frames and bees from three or more hives. When you do this you can mix the bees quite successfully without them fighting as the confusion gives them something else to focus on – but it does need to be from three or more hives.

Before you start stealing frames of bees from any colony find the resident queen and pop her in a cage while you carry out this operation. You do not want to accidentally put the queen in one of your mating nucs!

You can make the nucs on the same day you transfer the queen cell. Alternatively, some people like to do it a couple of days in advance but then you will need to check in the nuc before adding your queen cell to remove any emergency queen cells that they make.

Into your nuc place two frames of sealed brood and adhering bees. It doesn't matter if there is some unsealed brood and eggs on the frames but choose mainly sealed brood because these will soon be emerging to provide more bees to look after the new queen. Place a frame full of stores next to these two frames. Shake a couple more frames worth of bees into the nuc. These need to be young bees, which can be found looking after the brood in the centre of the brood nest, because the older foraging bees will fly straight back to their parent colony. Wedge your queen cell that you have cut from your Miller frame (a cocktail stick can help) between the two frames of brood and then blank off the rest of the space either by using a dummy board or a block of polystyrene. Close up the nuc box.

# Mini mating nucs

I have purposely not provided information about how to set up and use mini mating nucs. For those of you that are only looking to produce one or two queens the method described above will be suitable and you are likely to have all the necessary equipment. Using mini mating nucs requires more expense due to the additional equipment required and they also require some careful management to establish and maintain them throughout the process. This omission does not mean they are not an option, but this booklet is focussing on an accessible method that requires little extra equipment or expertise.

Once you become familiar with the process of raising new queens and maybe you are ready to increase the numbers of queens that you produce, you will probably find that mini mating nucs are worth the investment and worth taking the time to master.

# Back to the cell raiser colony

Briefly turn your attention back to your cell raiser colony. If you have removed all the queen cells to put into mating nucs you will need to reinstate the queen that you originally removed from this colony. Do this by re-joining the nuc that you put her in, to the queenless colony using the newspaper method of uniting. You may find that as the queen has been in that nuc for approximately three weeks that they have expanded, and you may have even needed to move the colony into a full size hive. This can still be joined to the queenless cell raiser colony using the newspaper method.

Fig 19. Unite the queen back to the cell raiser colony using the newspaper method.

Remember, you can re-queen the cell raiser colony by leaving one of the queen cells that they have raised on the Miller frame in situ. The queen can emerge and get mated and they will also draw comb in the space left where you cut the zig zag shape in the comb and very quickly it will return to looking like a regular frame. Now your only decision is to decide what you do with the queen in the nuc box that you originally removed from the cell raiser colony – that is for you to decide!

# The mating process

Where you position the mating nuc is not critical. It can stay in the apiary with your other hives or you may prefer to take it somewhere else. If you do transport it somewhere, be careful of that queen cell that you have just wedged into the frame. Leave it alone now for at least a week. The drones that mate with your virgin queens will come from apiaries in the surrounding area. They collect in drone congregation areas and the queens are able to find these locations and have the best chance of mating with a variety of drones. You do not need to worry about your queens mating with drones from your own hives as there are so many present from all the surrounding apiaries that this rarely happens.

After a week you will need to check that they have enough stores. Some care is required here because you do not want to interrupt a mating flight and risk interfering with the queen finding her way back into the colony when she returns from one of these flights. After the queen has emerged from her cell, she will spend a couple of days maturing and then when the weather is suitable for a mating flight she will go out in search of drones. These flights usually take place in the early afternoon but if it is a particularly warm day or the weather has been poor and a suitable window occurs then they may take place at other times of the day. Therefore, open the nuc in the late afternoon rather than during the middle of the day when there is a risk that the virgin queen might be out on a mating flight. While you have the nuc open check the queen cell to see if she has successfully emerged. If all looks well close up the nuc and leave for at least another week after which you can check if your new queen has started laying.

Once you have eggs in your nuc you can now leave them to build up a bit more by adding another frame of drawn comb and make sure they have sufficient stores. When you are happy that she has an even laying pattern and the brood looks healthy you can join this colony onto the hive you want to requeen using the newspaper method (remembering first to remove the old queen!). Alternatively, you can allow them to build up to a full size nuc adding more frames as they need it and when it is strong enough transfer them to a full size hive.

Fig 20. Allow the queen to continue laying and begin to build the colony. Look out for an even laying pattern and healthy brood.

# Final thoughts

The whole process will take approximately four weeks and it is incredibly rewarding to produce your own queens. There is nothing technically difficult about the process, it just requires patience and attention to detail. The important things to concentrate on are

▶ Select your best colony to take the donor larvae from

▶ Make the cut on the Miller frame where the larvae are less than 24 hours old

▶ Ensure the cell raiser colony is strong, healthy and well provisioned with pollen and nectar

▶ Handle the queen cells carefully when you transfer them to your mating nuc

▶ Remember to work out the correct day to transfer the queen cells to the mating nuc. You do not want them all to emerge in the cell raiser colony

▶ Be patient during the mating process. It could take anything from one to four weeks before you see eggs being laid.

And finally, please don't be put off if you are not successful. There are many points at which something could go wrong and this will not always be down to something you have done. When you are first starting out with queen rearing it is realistic to expect a success rate of 50% so aim to rear twice as many queens as you need to compensate for this. As you get more experienced the success rate could be as high as 80%. Don't view any losses as a failure on your part but instead think of it as being part of the learning experience. Try to work out what went wrong and where and have another go.